Collins

easy learning

Maths

Ages 9–11

Volume = 16 cm³

Sarah-Anne Fernandes

How to use this book

This book is for parents who want to work with their child at home to support and practise what is happening at school.

- Ask your child what maths they are doing at school and choose an appropriate topic. Tackle one topic at a time.

- Help with reading the instructions where necessary, and ensure that your child understands what to do.

- Help and encourage your child to check their own answers as they complete each activity. Discuss with your child what they have learnt.

- Let your child return to their favourite pages once they have been completed, to play the games and talk about the activities.

- Reward your child with plenty of praise and encouragement.

Special features

- **Games:** There is a game on each double page that reinforces the maths topic. Each game is for two players, unless otherwise indicated. Some of the games require a spinner, which is easily made using the circles printed on the games pages, a pencil and a paper clip. Gently flick the paper clip with your finger to make it spin.

- At the bottom of every page you will find footnotes that are **Parent's notes**. These are divided into '**What you need to know**', which explain the key maths idea, and '**Taking it further**', which suggest activities and encourage discussion with your child about what they have learnt. The words in bold are key words that you should focus on when talking to your child.

ACKNOWLEDGEMENTS

The author and publisher are grateful to the copyright holders for permission to use quoted materials and images.

p.4 © owatta/Shutterstock.com; p.7 © HitToon. Com/Shutterstock; p.9 © Shutterstock.com/rudall30; p.12 © Milen/Shutterstock.com; p.13 © © Shutterstock.com/HitToon.Com; p.15 © sahua d/Shutterstock.com; p.22 © Miguel Angel Salinas Salinas/Shutterstock.com; p.22 © 2009 Jupiterimages Corporation; p.28 © Clipart.com; p.38 © Anton Brand/Shutterstock.com; p.39 © peachbite/Shutterstock.com

Every effort has been made to trace copyright holders and obtain their permission for the use of copyright material. The author and publisher will gladly receive information enabling them to rectify any error or omission in subsequent editions. All facts are correct at time of going to press.

Published by Collins
An imprint of HarperCollins*Publishers*
1 London Bridge Street
London SE1 9GF

© HarperCollins*Publishers* Limited

ISBN 9780007559831

First published 2014

10 9 8 7 6 5 4 3

All rights reserved. No part of this publication may be reproduced, stored in a retrieval system, or transmitted, in any form or by any means, electronic, mechanical, photocopying, recording or otherwise, without the prior permission of Collins.

British Library Cataloguing in Publication Data.

A CIP record of this book is available from the British Library.

Publishing manager: Rebecca Skinner
Author: Sarah-Anne Fernandes (SolveMaths Ltd)
Assistant author: Gareth Fernandes (SolveMaths Ltd)
Commissioning and series editor: Charlotte Christensen

Project editor and manager: David Mantovani
Cover design: Susi Martin and Paul Oates
Inside concept design: Lodestone Publishing Limited and Paul Oates
Text design and layout: Q2A Media Services Pvt. Ltd
Artwork: Rachel Annie Bridgen, Q2A Media Services
Production: Robert Smith
Printed and bound by Printing Express Limited, Hong Kong

FSC MIX
Paper from responsible sources
FSC™ C007454

5 EASY WAYS TO ORDER

1. Available from www.collins.co.uk
2. Fax your order to 0844 576 8131
3. Phone us on 0844 576 8126
4. Email us at education@harpercollins.co.uk
5. Post your order to: Collins Education, Freepost GW2446, Glasgow G64 1BR

Contents

Numbers to 1000000

- Write the value of the digit 6 in each of these numbers.

624801 **564927** **936779** **801962** **799692** **101996**

- Write the value of the digit 7 in each of these measures.

271653 km   **137651 litres**   **6879 kg**   **£781995**   **31762 m**

Comparing and ordering

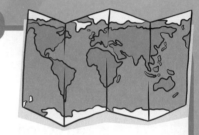

- Write these populations in order from the largest to the smallest.

European city	Athens	Lisbon	Milan	Paris	Zurich
Population	789010	474690	1305150	2211238	416759

Worldwide city	Abu Dhabi	Auckland	Delhi	San Francisco	Sao Paulo
Population	613368	1377541	9879120	825863	11390222

What you need to know At this stage your child is learning to recognise the **place value** of each **digit** in a number to at least 1000000. For example **987232** is made up of 900000 (9 **hundred thousands**), 80000 (8 **ten thousands**), 7000 (7 **thousands**), 200 (2 **hundreds**), 30 (3 **tens**) and 2 (2 **units** or 2 **ones**). This knowledge will help your child to compare and order numbers.

Game: Place value planets

You need: two paper clips, two pencils, a set of different coloured counters each.

- Take turns to spin both spinners.
- Find a planet that matches the place value of the digit spun and cover it with a counter.

Example: '6' and 'ten thousand' means 60 000. Cover the 864 310 planet because this number contains 60 000.

- If you can't cover a planet, or it has already been covered, miss a turn.
- The winner is the player with more planets covered.

ten thousand | hundred thousand

thousand

5 | 7

6

805 222

864 310

847 299

171 668

594 129

713 981

601 993

156 085

436 291

Missing numbers

- Fill in the missing number on each shape.

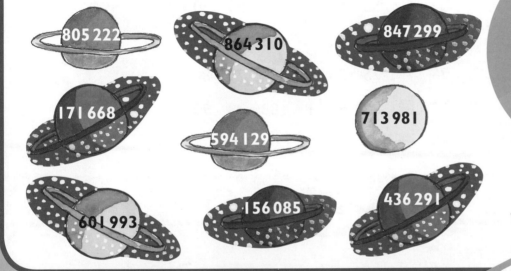

$965\,125 = 900\,000 + \boxed{} + \boxed{} + 100 + 20 + 5$

$864\,121 = 800\,000 + 60\,000 + 4000 + \bigcirc + 20 + \bigcirc$

$1\,175\,336 = \diamondsuit + \diamondsuit + 70\,000 + 5000 + 300 + \diamondsuit + 6$

Taking it further Choose two planets from the game above. Ask your child to read out the numbers. Look at the two numbers chosen and decide whether to use **less than** (<), **more than** (>) or **equal to** (=) between the two numbers.

Number properties

Prime numbers

- Circle the prime numbers on the 100 square.

1	2	3	4	5	6	7	8	9	10
11	12	13	14	15	16	17	18	19	20
21	22	23	24	25	26	27	28	29	30
31	32	33	34	35	36	37	38	39	40
41	42	43	44	45	46	47	48	49	50
51	52	53	54	55	56	57	58	59	60
61	62	63	64	65	66	67	68	69	70
71	72	73	74	75	76	77	78	79	80
81	82	83	84	85	86	87	88	89	90
91	92	93	94	95	96	97	98	99	100

Prime factors

- In the table below, write down all the factors for each number. Then write down which of these factors are prime factors. The first one has been done for you.

Number	All factors	Prime factors
6	1, 2, 3, 6	2, 3
20		
28		
34		

What you need to know   At this stage your child is learning the following key mathematics vocabulary:
- **Factor:** a factor of a number is a whole number that divides exactly into it. For example, factors of 12 are 1, 2, 3, 4, 6 and 12.
- **Multiple:** a multiple is the product when you multiply one whole number by another. For example, some multiples of 5 are 15, 20, 25, 30, 35 and 40.
- **Prime number:** a number that you can only divide by one and itself. For example, 7 is a prime number because it divides only by 7 and 1. Note that 1 is not a prime number!
- **Prime factor:** a factor that is also a prime number. For example, 2 and 3 are the only prime factors of 12.

Your child is also learning about **square numbers** and **cube numbers**.

Game: Square numbers

You need: a paper clip, a pencil, a set of different coloured counters each.

- Take turns to spin the spinner.

- Multiply the number shown on the spinner by itself. Place a counter on the square number in the honeycomb.

- The first player to get three in a row (horizontally or diagonally) wins!

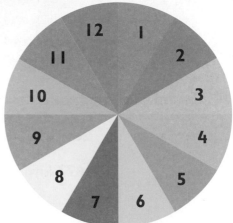

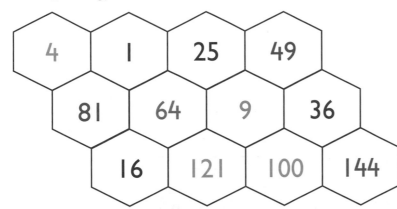

Odd one out

- In each row circle the number that is the odd one out, and explain why.

> **TIP:** Think about factors.

15   3   5   1   6   _____

27   9   40   3   1   _____

8   7   4   1   2   _____

Taking it further Look again at the game above. Check that all the numbers in the honeycomb are **square numbers**. Work through all the numbers from 1 to 12, multiplying each number by itself.
Ask your child if they know what a **cube number** is. Explain that a cube number is when you multiply any number by itself and then by itself again. For example, "5 cubed" (written as 5^3) is $5 \times 5 \times 5 = 25 \times 5 = 125$. Together work out 2^3, 3^3 and 4^3.

Rounding whole numbers

- Round the length of each river to the nearest 10 km, 100 km and 1000 km.

River	Length	10 km	100 km	1000 km
Nile	6672 km			
Mississippi	6275 km			
Ganges	2526 km			
Amazon	6489 km			
Congo	4749 km			

- Round the sum of money raised for each charity to the nearest £1000, £10 000 and £100 000.

Charity	Money raised	£1000	£10 000	£100 000
Age UK	£115 345			
Africa Wildlife	£592 751			
Homeless	£851 856			
Water Aid	£913 172			

What you need to know At this stage your child is learning to **round** any number up to 1 000 000 to the nearest 10, 100, 1000, 10 000 and 100 000. Encourage your child to remember to **round down** when the place value column is less than 5 and **round up** when it is 5 or more.
- 6642 rounded to the nearest 10 is 6640 because 2 ones (units) are less than 5 ones.
- 6642 rounded to the nearest 100 is 6600 because 4 tens are less than 5 tens.
- 6642 rounded to the nearest 1000 is 7000 because 6 hundreds are more than 5 hundreds.

Game: Round up!

You need: a paper clip, a pencil, paper.

- Take turns to spin the spinner five times. Write down the 5-digit number you make.

- Round the number to the nearest 10 000. If the number rounds up then score ten points. If it rounds down then score one point.

- Keep a running total of the points scored.

- The player who scores more points after five rounds wins!

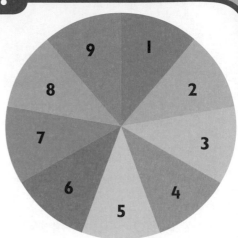

Jolly joke

When do 177 sheep become 200 sheep?

When you've rounded them up!

At the airport

1. Bawding Airport has estimated that the number of passengers arriving per day is 200 000 (to the nearest hundred thousand).

   a) What is the highest number of passengers that could have arrived per day?

   b) What is the lowest number of passengers that could have arrived per day?

2. Browns Airport Terminal 2 has estimated that it has 60 000 flights each year (to the nearest ten thousand).

   a) What is the greatest number of flights that could have arrived each year?

   b) What is the lowest number of flights that could have arrived each year?

Taking it further Ask your child to list ten numbers that will be 1 000 000 when rounded to the nearest 100 000. Encourage your child to think of numbers that must be rounded up as well as rounded down.
Can your child now think of an efficient way of writing all the possible numbers that can be rounded to 1 000 000?
Suggest that they use the symbols **less than** (<), **more than** (>) and **equal to** (=).

Number patterns

Counting in powers of ten

- Continue these number sequences.

1110	2110	3110			
34562	44562	54562			
131782	231782	331782			
567943	577943	587943			
867101	868101	869101			

Negative and positive numbers

- Fill in the missing numbers.

−12	−8		0	4		12
−15	−10		0			15
	−9	−7			−1	1
−10	−8	−6				2

What you need to know At this stage your child is learning to count forwards and backwards in **powers of ten** (i.e. counting in 10s, 100s, 1000s, 10000s, 100000s). They are also learning to count forwards and backwards with positive and negative numbers through zero.

Game: City temperatures

You need: two paper clips, two pencils, paper.

- Take turns to spin both spinners. Look at the table below. Find the difference in temperature between the two cities.

- The player with the larger difference in temperature scores five points.

- Play three rounds. The player with more points wins!

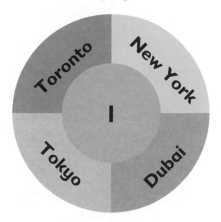

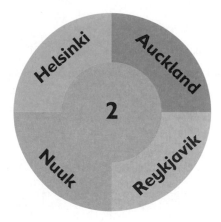

Toronto: 7°C	New York: 9°C	Tokyo: 4°C	Dubai: 13°C
Auckland: −2°C	Reykjavik: −7°C	Nuuk: −4°C	Helsinki: −3°C

Bank account balance

- Mark takes £10 from his bank account each day. He pays in a cheque of £45 every Wednesday.

  If his bank account balance on Saturday is £70 before he has withdrawn money, what will be his balance at the end of the following Friday?

  _____

  _____

  _____

Jolly joke

Why are powers like fish?

Because they're all indices (in the seas)!

Taking it further Roll a dice six times to make a 6-digit number. Ask your child to count either forwards or backwards in a power of ten (i.e. counting in 10s, 100s, 1000s, 10000s or 100000s). Give your child plenty of practice, making sure they are confident with crossing the boundary. For example, 187 234, 197 234, 207 234, 217 234.

Decimal numbers

Place value

- Write the value of the underlined digit in each of these decimal numbers.

1.2̲52 [ ]   2.78̲5 [ ]   9.34̲9 [ ]

1̲.002 [ ]   0.001̲ [ ]   3.5̲ [ ]

Rounding

Below are the results of the Javelin Throw Men's Final from the 2012 Olympics in London.

- Round the results to the nearest whole number and to one decimal place.

				Whole number	One decimal place
1.	Keshorn Walcott	Trinidad and Tobago	84.58		
2.	Oleksandr Pyatnytsya	Ukraine	84.51		
3.	Antti Ruuskanen	Finland	84.12		
4.	Vitezslav Vesely	Czech Republic	83.34		
5.	Tero Pitkamaki	Finland	82.80		
6.	Andreas Thorkildsen	Norway	82.63		
7.	Spiridon Lebesis	Greece	81.91		
8.	Tino Haber	Germany	81.21		

What you need to know At this stage your child is learning to recognise and use **decimals**. They are learning the value of digits in numbers with up to **three decimal places**, e.g. 21.764 = 20 (20 tens) + 1 (1 unit) + 0.7 (7 **tenths**) + 0.06 (6 **hundredths**) + 0.004 (4 **thousandths**). They will use this knowledge to help them order and compare numbers with the same number of decimal places up to three decimal places. They are also learning to **round decimals** to the **nearest whole number** and to **one decimal place**. For example, 3.72 rounded to the nearest whole number is 4 because 7 in the tenths column is 5 or more (so round up), while 3.72 rounded to one decimal place is 3.7 because the 2 in the hundredths column is less than 5 (so round down).

Game: Three in a row

You need: a paper clip, a different coloured pencil each.

- Take turns to spin the spinner.
- Write the symbol shown on the spinner in one of the yellow circles below to make a correct decimal sentence.
- The first player to get three in a row (vertically, horizontally or diagonally) wins!

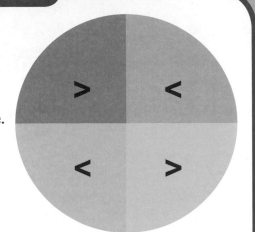

0.74 ◯ 0.76	5.67 ◯ 5.067	2.27 ◯ 2.027	0.72 ◯ 0.721
0.8 ◯ 0.08	0.769 ◯ 0.761	6.91 ◯ 9.4	9.023 ◯ 9.723
2.73 ◯ 2.74	0.101 ◯ 1.01	5.68 ◯ 5.61	9.111 ◯ 9.117
0.07 ◯ 1.7	0.1 ◯ 0.11	0.981 ◯ 0.989	1.3 ◯ 1.304

Decimal detective

Read the following statements.

> Round me to the nearest whole number. You get 4.
>
> Round me to the nearest tenth. You get 4.3.

- What number am I? Circle the number below.

4.21   4.28   4.23

- What other numbers could I be?

Jolly joke

Why didn't the line laugh?

He couldn't see the point.

Taking it further Ask your child to look at a shopping receipt with you. Ask them to round the price of each item to the nearest whole number in pounds.

Addition and subtraction

Adding large whole numbers

● Answer the following using column addition. The first one has been done for you.

$53\,129 + 1541 = \boxed{54\,670}$

```
    1
  5 3 1 2 9
+   1 5 4 1
  5 4 6 7 0
```

$46\,129 + 3245 = \boxed{}$

$6930 + 10\,701 = \boxed{}$

$17\,329 + 31\,456 = \boxed{}$

$623\,541 + 20\,193 = \boxed{}$

$989\,427 + 11\,092 = \boxed{}$

Subtracting large whole numbers

● Answer the following using column subtraction. The first one has been done for you.

$9399 - 5487 = \boxed{3912}$

```
  8 1
  9 3 9 9
- 5 4 8 7
  3 9 1 2
```

$7289 - 5396 = \boxed{}$

$13\,390 - 6458 = \boxed{}$

$22\,901 - 1681 = \boxed{}$

$72\,109 - 69\,251 = \boxed{}$

$101\,782 - 98\,643 = \boxed{}$

What you need to know   At this stage your child is learning to **add** and **subtract** whole numbers containing more than four digits using a formal **written method**. They are also learning to solve **multi-step problems**, deciding themselves how best to solve the problem.

Game: Football fans

You need: two paper clips, two pencils, paper.

Every Sunday there is an important football match at Wimbles Stadium. There are 80 000 seats in the stadium. Some seats are taken up by home supporters and others by away supporters.

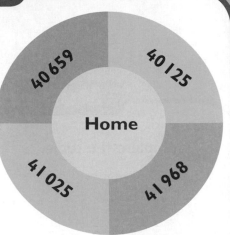

- Take turns to spin both spinners. Find the total of both home and away supporters attending the match. Then calculate how many empty seats are left at the stadium.

- The player with the lower number of empty seats wins three points. If the numbers are the same, both players get one point.

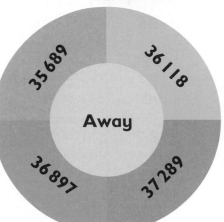

- Play five rounds. The player with the larger number of points wins!

Missing digits

- Fill in the missing digits.

$$
\begin{array}{r}
4\ \square\ 6\ 7 \\
+\ 5\ 6\ 4\ \square\ 0 \\
\hline
\square\ \square\ 9\ 7\ 7
\end{array}
$$

$$
\begin{array}{r}
9\ 8\ \square\ 6\ 7 \\
+\ 3\ 9\ 2\ 5\ 2 \\
\hline
\square\ \square\ \square\ 5\ \square\ \square
\end{array}
$$

$$
\begin{array}{r}
7\ 9\ 6\ 5\ 8 \\
-\ 5\ \square\ 8\ \square \\
\hline
\square\ \square\ 2\ \square\ 2
\end{array}
$$

Taking it further Ask your child real-life word problems involving at least two steps. For example: "The cinema sold 1082 film tickets during the week. On Saturday 394 tickets were sold and on Sunday 321 tickets were sold. How many tickets were sold altogether from Monday to Friday?"
Try to use a range of vocabulary for addition (e.g. total, sum, altogether, more than) and subtraction (e.g. less than, take away, difference between).

Multiplying and dividing by 10, 100 and 1000

Multiplying by 10, 100 and 1000

● Complete the table.

✗	10	100	1000
24			
35			
654			
901			
8.7			
4.2			
1.24			

Dividing by 10, 100 and 1000

● Complete the table.

÷	10	100	1000
327			
501			
69			
82			
134.62			
47.59			
4.24			

What you need to know   At this stage your child is learning to multiply and divide whole numbers and decimals by 10, 100 and 1000. Encourage your child to move the digits to the left when multiplying and to the right when dividing. Remind your child to put in the 0 as a **place value holder**.

Game: Conversion fact stars

You need: a paper clip, a pencil, a set of different coloured counters each.

The conversion fact stars show conversions between two measures.

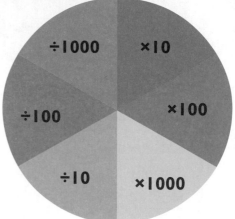

- Take turns to spin the spinner. Match the information spun to a fact star. Cover the star with a counter.

  Example: If you spin '×1000' then you can cover the fact star 'kg to g'.

- Miss a go if you can't cover a fact star.
- The player to cover the greater number of stars wins!

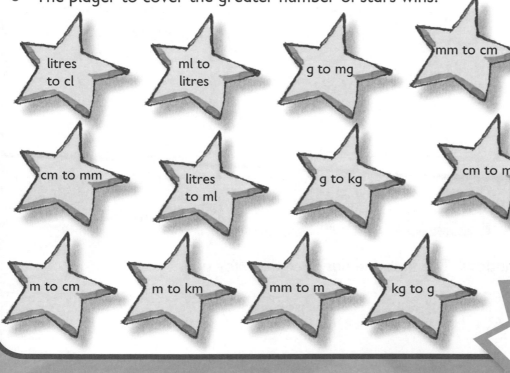

litres to cl

ml to litres

g to mg

mm to cm

cm to mm

litres to ml

g to kg

cm to m

km to m

m to cm

m to km

mm to m

kg to g

Maths chat

Tom and Tamara were having a discussion. Tom said that if you divide a 5-digit whole number by 1000 it will always give an answer to 3 decimal places.

- Is he correct? Yes ☐ No ☐
- Explain how you know.

_____

Taking it further Work through each of the 'conversion fact stars' above, asking your child to convert one measure to another. For example, if you choose 'mm to m' then you could ask your child "What is 8000 mm in m?". The answer will be 8 m: 1000 mm = 1 m so 8000 mm ÷ 1000 = 8 m.

17

Written multiplication

Multiplying whole numbers

- Answer these calculations. The first one has been done for you.

$321 \times 4 =$ [1284]

```
    3 2 1
×       4
  1 2 8 4
```

$1268 \times 3 =$ [ ]

$621 \times 13 =$ [ ]

$845 \times 27 =$ [ ]

$3521 \times 38 =$ [ ]

$5689 \times 19 =$ [ ]

Multiplying decimals

- Answer these calculations. The first one has been done for you.

$23.9 \times 4 =$ [95.6]

```
    1 3
  2 3.9
×     4
  9 5.6
```

$86.8 \times 7 =$ [ ]

$102.5 \times 8 =$ [ ]

$974.6 \times 6 =$ [ ]

$23.89 \times 3 =$ [ ]

$56.92 \times 5 =$ [ ]

What you need to know   At this stage your child is learning to **multiply** numbers with up to four digits by a 1-digit number or a 2-digit number using **short** or **long multiplication**.

Game: Floor plan

You need: a paper clip, a pencil, paper.

- Take turns to spin the spinner.

- Write down the name of the room shown on the spinner and then calculate the area of this room using the measurements on the floor plan.

- The first player to find the total area of all six rooms wins!

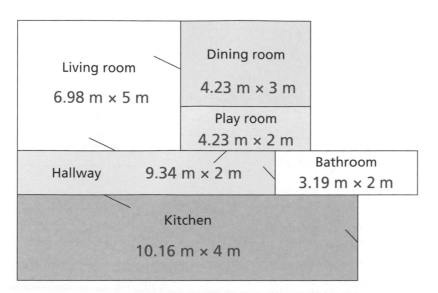

Living room
6.98 m × 5 m

Dining room
4.23 m × 3 m

Play room
4.23 m × 2 m

Hallway   9.34 m × 2 m

Bathroom
3.19 m × 2 m

Kitchen
10.16 m × 4 m

Jolly joke

What type of painting do maths teachers most enjoy?

Painting by numbers!

Multiplication puzzle

- Arrange the digits 9, 7, 6 and 4 into the boxes to give the **greatest** possible product.

 **.**

- Now arrange the digits 9, 7, 6 and 4 into the boxes to give the **smallest** possible product.

 **.**

Taking it further Ask your child to use rounding to check if their answers on page 18 are reasonable. For example, 321 × 4 rounded to the nearest 100 is 300 × 4 = 1200. Therefore, 1200 is a close estimation of what the answer will be.

Division

Dividing larger numbers

- For each of the following division questions, express the answer using a remainder, a fraction, and a decimal. The first one has been done for you.

$218 \div 8 =$ 27 r 2

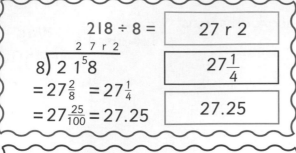

$= 27\frac{2}{8} = 27\frac{1}{4}$

$= 27\frac{25}{100} = 27.25$

$27\frac{1}{4}$

27.25

$171 \div 6 =$

$891 \div 5 =$

$8455 \div 4 =$

Division problems

1. It is Sarah's birthday and she brings into class a birthday log cake to share equally among her 8 friends. The cake is 30 cm long. How long must each slice be? Give your answer as a decimal.

2. Thomas has 850 grams of sugar, which he must divide equally between 6 cakes. How much sugar must he use for each cake? Express your answer as a fraction.

3. a) Cheryl has 192 seashells. She wants to share them between her 5 friends. How many shells does each friend get and how many shells will be left over?

   b) Does it make sense to calculate how many shells each friend gets as a fraction or decimal? Explain your answer.

What you need to know At this stage your child is learning to **divide** numbers up to four digits by a 1-digit whole number using **short division**. They are expected to **interpret the remainders** by expressing them as a **whole number remainder**, a **fraction** or a **decimal**. For example, $432 \div 5$ is 86 remainder 2 or $86\frac{2}{5}$ or 86.4.
They are also learning to answer division questions, thinking carefully whether to round up or down depending on the context of the question. For example: "Seven friends want to go to town. A taxi will seat five people. How many taxis should they hire?" The answer is two taxis, because the two remaining friends still need a taxi to get to town.

Game: Raffle prize draw

You need: two paper clips, two pencils, a set of different coloured counters each.

There is a raffle at the school fair. There are six prizes and ticket numbers.

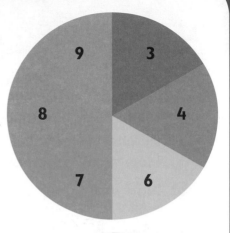

- Take turns to spin both spinners. Divide the ticket number for the prize by the number on the other spinner.

- If there is no remainder, you win the prize!

- Put one of your counters on that prize. The winner is the player with more prizes.

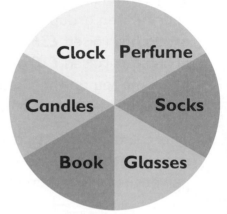

Division: Round up or round down?

1. A school is taking 113 children to a museum. The head teacher is going to hire some buses. Each bus has 78 seats. How many buses should she hire?

2. A farmer has 101 eggs to pack into boxes of 6. How many full boxes will he be able to sell?

3. There are 140 tomato plants in rows, with 6 plants in each row. How many complete rows of tomato plants are there?

Jolly joke

Teacher: Can you count from 1 to 20?
Pupil: I am not sure. How about if I count from 1 to 10 twice?!

Taking it further Ask your child to use rounding to check if their answers to the questions at the top of page 20 are reasonable. For example, 218 ÷ 8 rounded to the nearest 100 is 200 ÷ 8 = 25. Therefore, 25 is a close estimation of what the answer will be.

Problem solving

Money problems

1. James, Max and Michelle collect their one penny coins. Max has 569 penny coins. Michelle has 602 penny coins. James has double the number that Michelle has.

   a) How many penny coins do they have altogether?

   b) Now write your answer in pounds and pence.

2. Tiny Swim Fins are doing a swimathon for charity. Red group has raised £225, Green group has raised £367 and Orange group has raised £109. How much more money do they need to raise to reach their target of £1000?

3. There is a class party and I have been asked to buy some crisps. I have £5 to spend. The crisps cost 35p per packet. How many packets of crisps can I buy?

4. I buy 2.5 kg of potatoes. They cost 88p per kg. How much do I pay?

Scale drawings

- Draw each of the following objects to scale on the 1 cm squared paper. Use a scale of 1 cm = 100 cm.

    Bus: length = 10 m, height = 5 m   Car: length = 3.5 m, height = 1.5 m

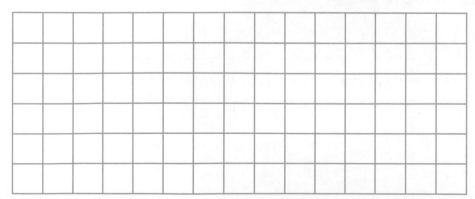

What you need to know   At this stage your child is learning to solve a **range of problems** in a variety of contexts. They are expected to use all four operations to solve problems in the context of money. They are also learning to use their knowledge of multiplying by 10, 100, 1000 and 10000 to complete scale drawings.

Game: Decimal totals

You need: a pencil.

- Move through the maze from start to finish by adding numbers that will give you the finish number. You may move across, down, up or diagonally. Mark your route with arrows.

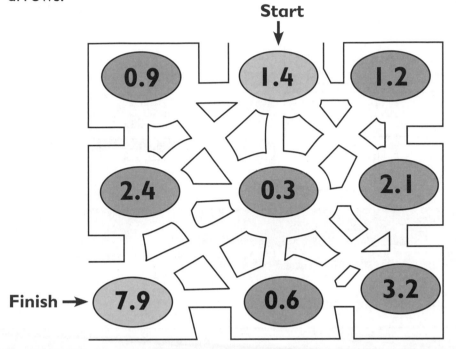

Start

0.9   1.4   1.2

2.4   0.3   2.1

3.2

Finish → 7.9   0.6

Jolly joke

What goes up but never goes down?

Your age!

Age puzzle

- Work out this age puzzle.

Last year, my age was a square number.

Seven years ago, my age was a prime number.

Next year, my age will be a cube number.

How old am I?

Happy Birthday

Taking it further   Ask your child to think of decimal pairs that total 1. For example 0.6 + 0.4 = 1; 0.82 + 0.18 = 1. Ask your child to find how many different pairs they can make.

Fractions

Equivalent fractions

- On each row, circle the fractions that are equivalent to the fraction in red. You can use the fraction wall to help you.

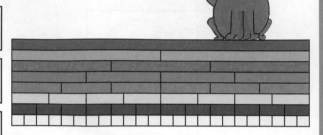

$\frac{1}{2}$	$\frac{2}{2}$	$\frac{8}{16}$	$\frac{3}{6}$	$\frac{5}{8}$	$\frac{5}{10}$
$\frac{1}{3}$	$\frac{3}{3}$	$\frac{7}{21}$	$\frac{4}{12}$	$\frac{1}{10}$	$\frac{5}{15}$
$\frac{2}{5}$	$\frac{5}{10}$	$\frac{4}{10}$	$\frac{4}{8}$	$\frac{8}{20}$	$\frac{6}{15}$

Ordering fractions

- Order each set of fractions from the smallest to the largest.

$\frac{1}{5}$	$\frac{4}{10}$	$\frac{8}{10}$	$\frac{6}{20}$	$\frac{9}{10}$

$\frac{1}{4}$	$\frac{3}{8}$	$\frac{5}{16}$	$\frac{10}{16}$	$\frac{3}{4}$

$\frac{2}{3}$	$\frac{5}{12}$	$\frac{6}{12}$	$\frac{1}{3}$	$\frac{3}{24}$

What you need to know At this stage your child is learning to write **equivalent fractions** of a given fraction, for example $\frac{1}{5}$ is equivalent to $\frac{2}{10}$. They are also learning to **order fractions** when the denominators are not all the same. For example, order $\frac{1}{3}$, $\frac{1}{6}$, $\frac{5}{6}$ and $\frac{2}{3}$ from smallest to largest. First, find **a common denominator** for all the fractions. In this case it is 6. Then convert the fractions so $\frac{1}{3} = \frac{2}{6}$ and $\frac{2}{3} = \frac{4}{6}$. Therefore the order of fractions is $\frac{1}{6}$, $\frac{1}{3}$, $\frac{2}{3}$ and $\frac{5}{6}$.
Your child is also learning to convert **improper fractions** (fractions where the numerator is greater than the denominator), e.g. $\frac{5}{3}$, to **mixed fractions**. In this case, $\frac{5}{3}$ is $1\frac{2}{3}$ as it is made up of 1 whole $\left(\frac{3}{3}\right)$ and $\frac{2}{3}$.

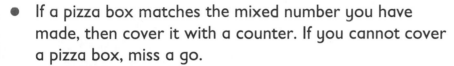
Game: Pizza boxes

You need: two paper clips, two pencils, a set of different coloured counters each.

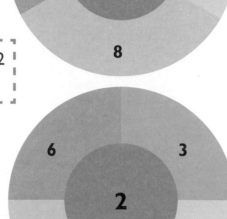

- Take turns to spin both spinners to make an improper fraction. The number from spinner 1 will be the numerator and the number from spinner 2 will be the denominator. Convert the improper fraction to a mixed number.

  > **Example:** If you spin '9' on spinner 1 and '5' on spinner 2 then the fraction will be $\frac{9}{5}$. $\frac{9}{5}$ is equal to $1\frac{4}{5}$.

- If a pizza box matches the mixed number you have made, then cover it with a counter. If you cannot cover a pizza box, miss a go.

- Keep taking turns. The player with more pizza boxes at the end of the game wins!

$2\frac{1}{3}$ $1\frac{3}{4}$ $1\frac{2}{5}$ $1\frac{1}{6}$ $2\frac{1}{4}$

$1\frac{4}{5}$ $1\frac{3}{6}$ $2\frac{2}{3}$ $1\frac{3}{5}$ $1\frac{2}{6}$

Missing box fraction puzzle

- Write down four different ways to make the fraction sentence correct.

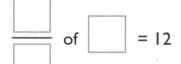 of 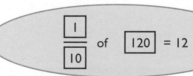 = 12

$\frac{1}{10}$ of 120 = 12

Taking it further Ask your child what the equivalent decimals of the following common fractions are: $\frac{1}{2}, \frac{1}{4}, \frac{3}{4}, \frac{1}{10}, \frac{1}{5}$. Explain that if they know $\frac{1}{10}$ then they can work out $\frac{2}{10}, \frac{3}{10}, \frac{4}{10}, \frac{5}{10}$ and so on.

More fractions

- Answer these questions. If appropriate, write the answer as a mixed number.

$\frac{4}{8} + \frac{7}{8} =$ [ ]

$\frac{1}{3} + \frac{4}{12} =$ [ ]

$\frac{2}{5} + \frac{10}{15} =$ [ ]

$\frac{7}{10} - \frac{4}{10} =$ [ ]

$\frac{3}{4} - \frac{5}{8} =$ [ ]

$\frac{4}{5} - \frac{3}{10} =$ [ ]

Multiplying fractions

- Answer these questions. The first one has been done for you.

$\frac{1}{2} \times 3 =$ $\boxed{\frac{3}{2} = 1\frac{1}{2}}$

$\frac{1}{2} \times 5 =$ [ ]

$1\frac{1}{3} \times 2 =$ [ ]

$\frac{1}{4} \times 7 =$ [ ]

$1\frac{4}{5} \times 3 =$ [ ]

$\frac{1}{5} \times 6 =$ [ ]

What you need to know At this stage your child is learning to **add** and **subtract fractions** that do not always have the same **denominator**, e.g. $\frac{1}{6} + \frac{1}{12}$ or $\frac{1}{6} - \frac{1}{12}$. In this case, your child has to find a **common denominator**. In this example it will be 12, so convert $\frac{1}{6}$ to 12ths: $\frac{1}{6} = \frac{2}{12}$. Therefore, $\frac{2}{12} + \frac{1}{12} = \frac{3}{12} \left(= \frac{1}{4}\right)$ and $\frac{2}{12} - \frac{1}{12} = \frac{1}{12}$.

Your child is also learning to **multiply proper fractions and mixed numbers** by whole numbers, e.g. $\frac{1}{2} \times 3 = \frac{3}{2} = 1\frac{1}{2}$.

Game: Fraction wheel spinners!

You need: two paper clips, two pencils, paper, four counters each.

- Each player should choose a fraction wheel (green or blue).

- Take turns to spin your fraction wheel.

- Subtract the fraction in the yellow centre circle from the fraction spun in the red inner circle. Check your answer with your opponent. If they agree it is correct, then place a counter on the corresponding outer ring.

- The winner is the player to cover all four of their outer rings first.

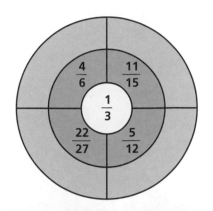 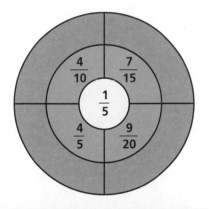

Jolly joke

How far should you open a window during a maths lesson?
Just a fraction!

Fraction magic squares

- Complete these magic squares.
  Each row, column and diagonal must add up to the same number.

The magic number is $\frac{15}{6}$.

		$\frac{8}{6}$
$\frac{9}{6}$		
	$\frac{7}{6}$	1

The magic number is ☐

$\frac{3}{4}$	$2\frac{1}{2}$	2
	1	

TIP: Find the value of one row first to find the magic square value.

Taking it further   Look again at the magic squares. Ask your child to convert the improper fractions in the first magic square to mixed numbers, e.g. $\frac{9}{6} = 1\frac{3}{6} = 1\frac{1}{2}$. For the second magic square, ask your child to convert the mixed numbers in the magic square to equivalent decimals, e.g. $2\frac{1}{2} = 2.5$.

Percentages

Here are some children's test scores. Each test is out of 100 marks.

- Write each test score as a percentage.

	Maths score out of 100	Percentage (%)
Annabel	54	
Sanjay	87	
Richard	72	
Catherine	63	
Harry	91	

Decimal and percentage equivalents

- Draw a line to match each decimal to the correct percentage.

20%   80%   50%   70%   10%   100%   25%

0.25   0.7   1.0   0.8   0.1   0.5   0.2

What you need to know At this stage your child is learning to recognise the symbol for **per cent** (%) and understand that **'per cent'** means 'out of 100'. Your child is learning to write percentages as a **fraction out of 100**, e.g. 54% is $\frac{54}{100}$. You can also write percentages as decimal numbers, e.g. 54% is 0.54. Your child is also developing their understanding of fraction, decimal and percentage equivalents, e.g. $\frac{1}{10}$ = 0.1 = 10%.

Game: Sale tags!

You need: a paper clip, a pencil, a set of different coloured counters each.

- Take turns to spin the spinner. Find the sale price tag that is equivalent to the fraction shown on the spinner. Cover the sale price tag with a counter.

- If the sale price tag has already been covered then you miss a go.

- After five rounds, the player who has more sale tags covered wins!

Spinner: $\frac{3}{10}$, $\frac{1}{2}$, $\frac{1}{4}$, $\frac{1}{10}$, $\frac{3}{4}$, $\frac{1}{5}$, $\frac{2}{5}$, $\frac{6}{10}$, $\frac{33}{100}$

20% OFF   **40% OFF**   **30% OFF**

75% OFF   **33% OFF**   **50% OFF**

10% OFF   **60% OFF**   **25% OFF**

Jolly joke

What is the most mathematical part of speech?

The add verb!

Discount prices

- Find the new price of each item.

10% OFF £90

20% OFF £120

25% OFF £60

Taking it further   Point out sales discounts that you may see when in shops or when shopping online. Ask your child to calculate the reduced price. If they have to work out a more complex percentage, e.g. 15%, encourage them to calculate 10% first, then 5%, and then add them together.

Area and perimeter

- Find the perimeter of each pond.

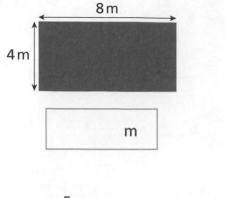

8 m

4 m

m

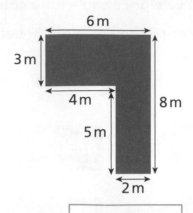

6 m

3 m

4 m

8 m

5 m

2 m

m

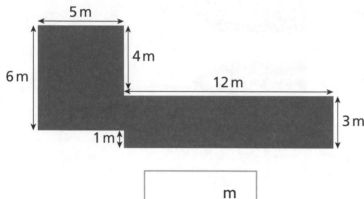

5 m

4 m

6 m

12 m

3 m

1 m

m

Area of irregular polygons

- Count the squares to find the area of each flower bed.
  Each square is equal to 1 m^2.

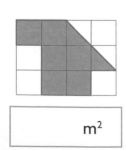

m^2

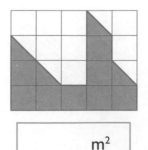

m^2

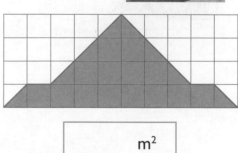

m^2

What you need to know At this stage your child is learning to **measure** and calculate the **perimeter** of rectangular shapes in centimetres and metres where there are missing lengths. They are also learning to estimate the **area** of irregular shapes by counting the number of squares and to find the area of rectangles using **cm^2** and **m^2**.

Game: Largest area!

You need: a 1–6 dice, plain paper, a ruler, a pencil.

- Take turns to roll the dice twice. Draw a rectangle accurately, using a ruler, using the two numbers you rolled.

> **Example:** If you roll 4 and 3 you can draw this rectangle:
>
> 4 cm
>
> 3 cm

- Find the area of your rectangle.
- The player with the greater area wins a point.
- Play five rounds. The person with more points wins!

Jolly joke

Why did the student eat his maths exam?

Because the teacher said that it was 'a piece of cake'!

Missing lengths perimeter puzzle

- Use the information given to work out the perimeter for each of these shapes.

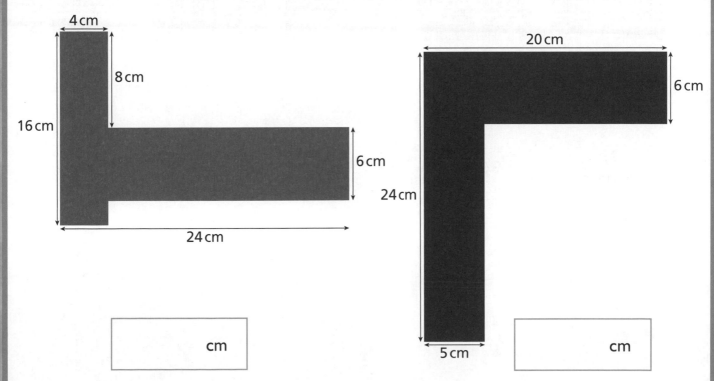

4 cm

8 cm

16 cm

6 cm

24 cm

20 cm

6 cm

24 cm

5 cm

cm

cm

Note: Images are not drawn to scale.

Taking it further   Challenge your child to think about how they might work out the **area** of the T-shape and L-shape above. Encourage your child to see that both shapes are made up of two rectangles. They can then use their knowledge of working out the area of rectangles to give the answer.

Measures

- Convert the following measurements.

5 litres = [　　　] ml   1700 ml = [　　　] litres   9.9 litres = [　　　] ml

7.8 kg = [　　　] g   2400 mg = [　　　] g   6072 g = [　　　] kg

756 cm = [　　　] m   1210 m = [　　　] km   703 m = [　　　] cm

Imperial and metric equivalents

- Draw a line to match each imperial unit to its metric equivalent.

| 1 inch | 1 foot | 1 ounce | 2.2 pounds | $1\frac{3}{4}$ pints |

| 28 grams | 30 cm | 2.5 cm | 1 kg | 1 litre |

What you need to know At this stage your child is learning to **convert** between **different units of measure**, e.g. kg to g, km to m, m to cm, cm to mm, litres to ml. This will build on their knowledge of multiplying and dividing by 10, 100 and 1000. They are also learning to understand and use equivalences between **metric** units (e.g. m, cm, litres, ml, kg, g) and **imperial units** such as inches, pounds and pints.

Game: Measures maze

You need: a coloured pencil. This game is for one player only.

- Colour in the route through the maze grid from the **START** box to the **FINISH** box.
- You can only move to a box that contains a **reasonable** measure in it.
- You can move left or right, up or down but not diagonally.

START	A small apple weighs 100 grams	The average capacity of a standard size kettle is 4 litres	A small carton of orange juice contains 1.2 litres
A pen weighs 70 grams	An average bath tub holds 270 litres	The average height of a door is 1.96 metres	A fuel can holds 4.5 litres of fuel
The length of a bus is 25 m	The length of a standard single bed is 400 centimetres	An average car is 20 metres long	The height of a cereal box is 25 centimetres
A teaspoon can hold about 45 millilitres of liquid	An egg weighs 0.5 kg	A feather weighs 200 grams	FINISH

Water jug

Look at the water jug.

- How much more water needs to be poured into this jug to make $\frac{9}{10}$ litre?

- How much water needs to be poured out of this jug to leave $\frac{1}{4}$ litre?

Jolly joke

What is one of the longest words?

Smiles: the first and last letters are separated by a mile!

Taking it further   Look at different items of packaging. Are the measurements given in metric units or imperial units? Ask your child where they might see imperial units being used in everyday life. Together try to find some real-life examples, perhaps around the home (e.g. in an old recipe book) and when out (e.g. road signs in miles).

Volume

Volume of solid shapes

- Find the volume of these solid shapes by counting how many 1 cm³ blocks there are in each one.

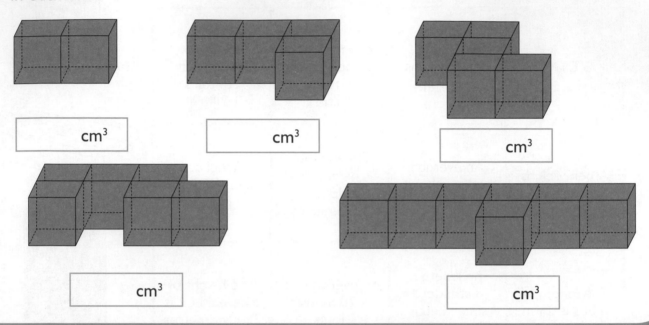

[ ] cm³	[ ] cm³	[ ] cm³

[ ] cm³	[ ] cm³

Volume of cuboids

- Find the volume of each of these cuboids by counting how many 1 cm³ blocks there are.

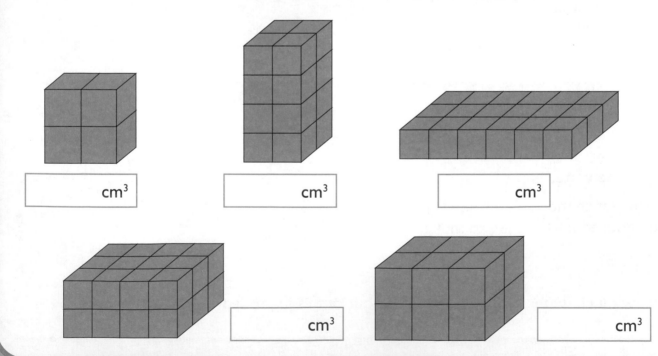

[ ] cm³	[ ] cm³	[ ] cm³

[ ] cm³	[ ] cm³

Game: Fill the bottle

You need: a paper clip, paper, a green colouring pencil, a purple colouring pencil.

- Each player should choose a bottle (purple or green).

- Take turns to spin the spinner. Complete the calculation on the spinner and use the answer as the amount of liquid to add to your bottle.

- The first player to fill their bottle wins!

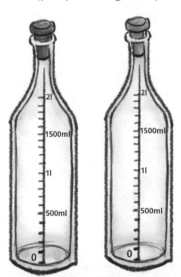

$\frac{3}{4}$ of a litre

10% of a litre

$\frac{1}{4}$ of a litre

$\frac{3}{10}$ of a litre

$\frac{1}{2}$ a litre

Jolly joke

Why did the cube laugh?

Because he saw the square dancing!

Cuboid puzzle

A cuboid is made up of 18 centimetre cubes. 1 cm³ =

- Find three different ways to make up a cuboid with 18 centimetre cubes. An example has been given for you.

Length	Width	Height	Drawing
9 cubes	2 cubes	1 cube	

Taking it further Look at different containers with your child, e.g. long tall jugs, short wide jars. Ask your child to predict the capacity of each container. Which container do they think will hold the most and which the least? Test their predictions by filling each container with water and then measuring how much each container holds.

Time

- You have to check in at the airport 2 hours and 20 minutes before departure time. Write the check-in time for each flight.

Flight number	Destination	Departure time	Check-in time
VP 4562	Paris	10:45	
VM 8796	Moscow	12:30	
VNY 9098	New York	13:55	
VL 4312	London	14:22	
VM 8543	Madrid	15:17	
VCT 4129	Cape Town	16:12	

Reading timetables

The timetable shows a day at the FunPark Activity Centre.

	Swimming	Breakfast	Archery	Football	Lunch	Chess	Hockey
Start	07:45	08:45	09:20	10:25	12:15	13:30	15:00
Finish	08:30	09:15	10:05	12:00	13:15	14:45	16:30

- What time did swimming start?

- What time did chess finish?

- How long was the archery class?

- What was happening at 09:00?

- What was the longest activity?

- What was the shortest activity?

- How long was the whole activity day?

What you need to know At this stage your child is learning to **convert time** between **different units**, e.g. hours to minutes, minutes to seconds, years to months, and weeks to days. They are also learning to read **timetables** confidently.

Game: Guided tours

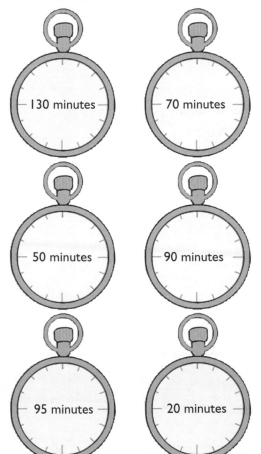

You need: a paper clip, a pencil, a set of different coloured counters each.

- Take turns to spin the spinner. On the timetable find the tour that matches the spinner.
- Calculate the length of the tour. Place a counter on the stopwatch that matches the duration of the tour. If you can't cover a stopwatch, miss a turn.
- The player who covers more stopwatches wins!

Spinner labels: Dungeon, Museum, National Park, Cave, Castle, Cathedral

Stopwatches: 130 minutes, 70 minutes, 50 minutes, 90 minutes, 95 minutes, 20 minutes

Tour	Start	Finish
Museum	09:00	10:30
National Park	11:15	12:25
Cave	12:30	12:50
Castle	13:00	15:10
Cathedral	15:30	17:05
Dungeon	17:25	18:15

Jolly joke
How can you make time fly?
Throw a clock out the window!

Visiting times

Class 4 are timetabled to visit the school library between 1:30 p.m. and 3 o'clock every Thursday afternoon.

- Give three different start and end times if they are allowed to visit for 45 minutes.

Taking it further Give your child lots of practice with reading timetables. Look at train timetables, which can be found online, and ask questions such as "What time does the next train leave?", "What is the latest train we can catch if we need to get to town for 10:30 a.m.?", and "How long will we wait at the train station if we arrive at 9:15 a.m.?".

More problem solving

Mixed word problems

Answer these problems.

1. A box contains 6 red pencils, 8 black pencils and 14 blue pencils. What fraction of the pencils are red?

2. Gavin and his sister went camping in France for 5 days. They then went to visit their aunt in Spain for 11 days. How many weeks and days were they away for?

3. The cost of a sofa is $\frac{2}{10}$ of the price of a new car. The new car costs £6542. How much does the sofa cost?

4. Sarah bought a watch and a ring. Each item cost £90. She later sold the watch for a 20% profit and the ring for a 10% loss.

   a) How much money did she get for the watch?

   b) How much money did she get for the ring?

   c) How much money did she make altogether?

Roman numeral puzzle

- Use Roman numerals to write the dates for these Olympic Games.

2012, London	
2004, Athens	
1996, Atlanta	
1988, Seoul	
1976, Montreal	

I = 1

V = 5

X = 10

L = 50

C = 100

D = 500

M = 1000

What you need to know   At this stage your child is learning to solve a **range of problems** in a variety of contexts. They are expected to **choose** the appropriate operation (i.e. **add**, **subtract**, **multiply** or **divide**). Your child is also learning to read **Roman numerals** to 1000 (M) and recognise years written in Roman numerals.

Game: Making word problems

You need: a paper clip, a pencil, paper.

- Spin the blue 'context' spinner once and the red 'operations' spinner twice.
- Use the information from the spinners to make up word problems.
- Solve the problems together. In this game, you both win!

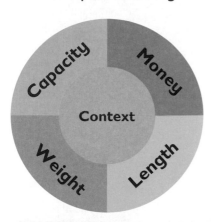 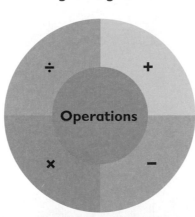

Example: If you spin **Money** on the 'context' spinner and **+** and **−** on the 'operations' spinner, you could make up the problem: "Jill went to the ice cream parlour with two friends. They bought an ice cream sundae for £2.65, a milkshake for £3.50 and an ice cream waffle for £4.25. How much change did they get from £20?"

Jolly joke

What has a hundred heads and a hundred tails?

One hundred pennies!

Fruit smoothies

The recipe shows the ingredients needed for a summer berry smoothie for 4 people.

250 ml plain yoghurt	350 g strawberries
400 g blueberries	20 ml honey
100 g raspberries	

- Write down the quantities needed for different numbers of people.

For 2 people

For 8 people

For 6 people

Taking it further Ask your child to look at the smoothie recipe, which is for four people. Can they work out the quantities for one person? How did they do this?

39

Angles

- Use a protractor to measure each of these angles. For each angle write what type of angle it is and give the measurement in degrees. An example has been done for you.

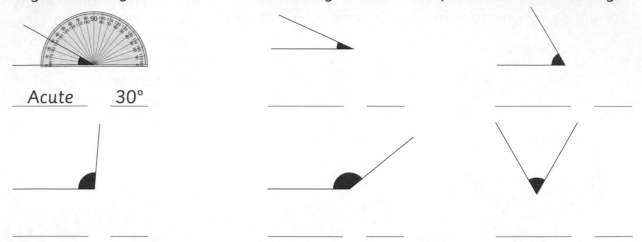

Acute   30°

_____   _____   _____   _____

_____   _____   _____   _____

Drawing angles

- Use the instructions below to draw two different angles.

1. Draw a straight line and place a dot at the left-hand end of the straight line. This will be the vertex of the angle.
2. Place the centre of your protractor on the vertex.
3. Find the required angle on the protractor (reading from 0 on the right-hand side of the protractor) and mark a small dot.
4. Join the small dot to the vertex with another straight line.

35° angle	110° angle

What you need to know At this stage your child is learning to use a **protractor** to draw a given angle and to measure angles to the nearest **degree**. They are also learning to identify **angles at a point**. They will know that a **whole turn** adds up to **360°** and that angles on a **straight line (half a turn)** add up to **180°**. They can use these facts to help them find missing angles.

Game: Ice skater pivots

You need: a paper clip, a protractor, a pencil, paper.

The ice skater is trying to do a full pivot turn. With practice she gets better each time. Measure her first pivot turn with a protractor.

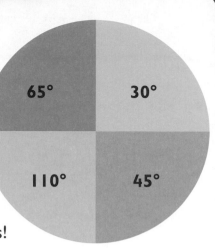

- Take turns to spin the spinner and add the amount to her first pivot turn.

- Each time you spin the spinner, add the amount to your previous pivot turn total.

- The first player to reach or go beyond a full pivot turn wins!

First pivot turn

Missing angles

- Work out the value of *x*, **without using a protractor**.

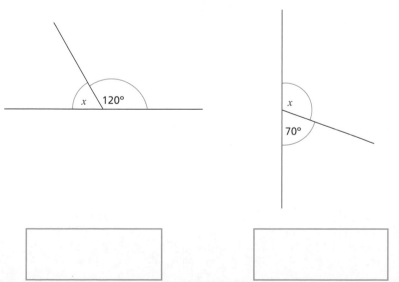

Taking it further Continue to practise finding missing angles without using a protractor. Ask your child to find the missing angle in a whole turn. Ask questions such as: "The ice skater has turned 243°, how much more does she need to turn to make a full turn?"

Shapes

- Tick the nets that will form a cuboid.

Other 3D shape nets

- Draw a line to match each red 3D shape to its correct blue net.

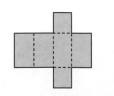   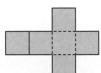

What you need to know At this stage your child is learning to identify 3D shapes from their 2D net representations. They are also learning to identify regular and irregular polygons. A **regular polygon** is a 2D shape where all the sides are the same length and all the angles are equal in size. An **irregular polygon** is a 2D shape where the lengths of the sides and sizes of the angles are not all equal.

Game: Regular or irregular 2D shapes

You need: two paper clips, two pencils, a set of different coloured counters each.

- Take turns to spin both spinners. Cover a shape to match what is shown on the spinners.

  Example: If you spin 'regular' on spinner 1 and 'hexagon' on spinner 2, then, with a counter, cover a hexagon that has six equal sides and six equal angles.

- The first player to cover three shapes in a row (vertical, horizontal or diagonal) wins!

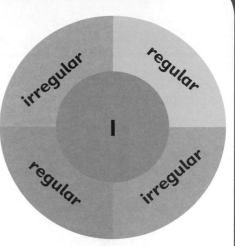

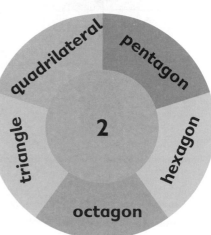

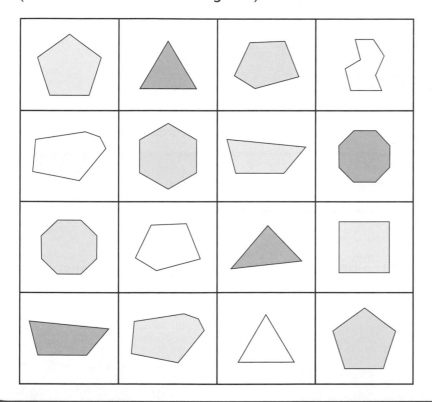

Jolly joke

What would you say if someone's parrot had died?

Poly...gon!

Cube puzzle

Here is the net of a cube.

- Draw two more different nets that make cubes.

Taking it further Ask your child to look for different items of packaging around the home. Together unfold the packaging to explore the net of the 3D shape.

Position and direction

● Tick the shapes that are a correct reflection in the mirror line.

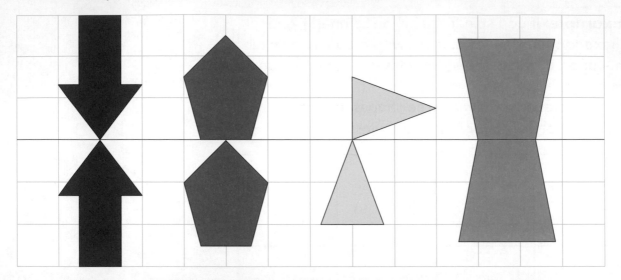

Translations

● The grid shows two shapes.
Explain how you would translate the orange shapes to cover the blue shapes.

Shape A: _____   Shape B: _____

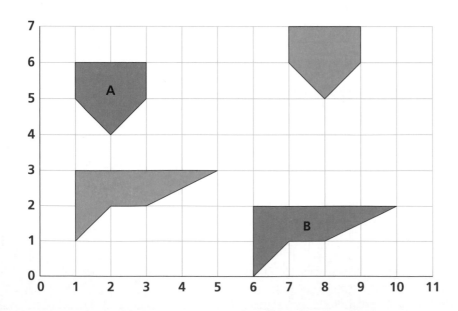

Tip: Use the following
language to help: up,
down, left, right.

What you need to know   At this stage your child is able to identify a shape after it has been **reflected** or **translated**.
They are continuing to describe position on a 2D grid using **coordinates**, e.g. (1, 5).

Game: Places in town

You need: a paper clip, a different coloured pencil each.

- Take turns to spin the spinner twice and make a coordinate.

  Example: If you spin '4' and '5', you can make the coordinates (4, 5) or (5, 4).

- If your coordinates match a place on the grid, then cross it off the grid with your coloured pencil.

- The first player to cross off five different places on the grid wins!

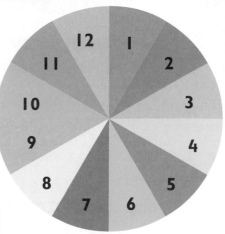

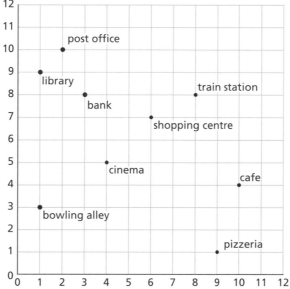

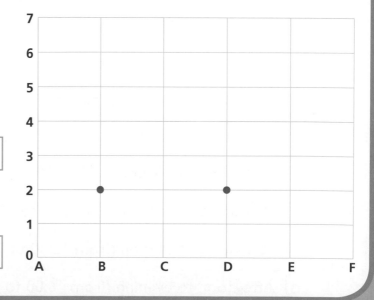

Jolly joke

Where are mathematicians buried?

The Symmetry!

Triangle coordinate puzzle

- (B, 2) and (D, 2) are two vertices of a triangle. Where could the third vertex be to form the following triangles on the grid?

Right-angled triangle

Isosceles triangle

Taking it further Using the triangle coordinate puzzle above, ask your child where the third vertex could be if they had to make a scalene triangle. Move on to ask your child to think where the third and fourth vertices could be in order to make a quadrilateral on the grid. How many different quadrilaterals can they make?

Line graphs

The table shows the temperature recorded every two hours during a day in December in the city of Timbuktu in the Sahara desert.

- Use the data in the table to construct a line graph.

Time of day	00:00	02:00	04:00	06:00	08:00	10:00	12:00	14:00	16:00	18:00	20:00	22:00	00:00
Temperature (°C)	5	2	6	11	24	35	41	39	33	28	19	11	5

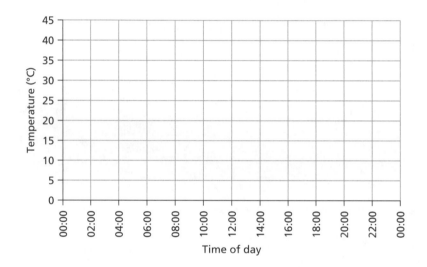

- Use the line graph you have drawn to answer these questions.

   1. At what time of day was the temperature the highest?

   2. At what time of day was the temperature the lowest?

   3. What was the change in temperature between 08:00 and 16:00?

   4. What was the change in temperature during each part of the day?

      a) Early morning (from 02:00 to 08:00)

      b) Morning to midday (from 08:00 to 12:00)

      c) Midday to afternoon (from 12:00 to 16:00)

      d) Afternoon to evening (from 16:00 to 20:00)

What you need to know   At this stage your child is learning how to construct **line graphs**. They are also learning to solve problems and use the information presented in line graphs.

You need: a 1–6 dice, a different coloured pencil each.

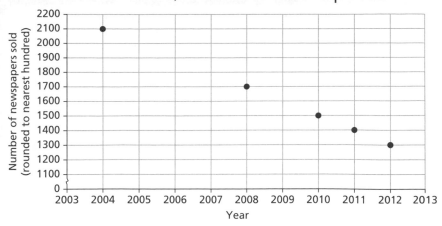

The line graph has some points missing.

- Take turns to roll the dice and match the number rolled to a clue. Tick the clue with your coloured pencil.

- If you roll the dice and you have already ticked the matching clue then miss a turn.

- The first player to tick all the clues wins!

- Now, work together using all the clues to complete the graph.

⚀	**Sales in 2003 were 1000 more than in 2013**
⚁	**Sales in 2013 decreased by 100 compared to 2012**
⚂	**Sales in 2005 were 100 more than in 2006**
⚃	**2000 newspapers were sold in 2006**
⚄	**Sales in 2007 were 100 less than in 2006**
⚅	**Sales in 2009 were the same as in 2008**

Jolly joke

If it takes six men one hour to dig a hole, how long would it take one man to dig half a hole?

You can't dig half a hole!

Explaining the data

- Suggest a reason why newspaper sales have decreased over the last ten years.

_____

- What product can you think of that may have had an increase in sales over the last ten years?

_____

Taking it further Ask your child to record the temperature of a cooling cup of tea (they will need a thermometer to do this). Ask your child to use this information to construct a line graph. Once they have checked their line graph with you, ask them to write six questions about their line graph for somebody to answer.

Answers

Pages 4–5

Place value

600 000 (six hundred thousand), 60 000 (sixty thousand), 6000 (six thousand), 60 (sixty), 600 (six hundred), 6 (six units)

70 000 km (seventy thousand kilometres), 7000 litres (seven thousand litres), 70 kg (seventy kilograms), £700 000 (seven hundred thousand pounds), 700 m (seven hundred metres)

Comparing and ordering

Paris, Milan, Athens, Lisbon, Zurich

Sao Paulo, Delhi, Auckland, San Francisco, Abu Dhabi

Missing numbers

60 000, 5000
100, 1
1 000 000, 100 000, 30

Pages 6–7

Prime numbers

1	②	③	4	⑤	6	⑦	8	9	10
⑪	12	⑬	14	15	16	⑰	18	⑲	20
21	22	㉓	24	25	26	27	28	㉙	30
㉛	32	33	34	35	36	㊲	38	39	40
㊶	42	㊸	44	45	46	㊼	48	49	50
51	52	(53)	54	55	56	57	58	(59)	60
(61)	62	63	64	65	66	(67)	68	69	70
(71)	72	(73)	74	75	76	77	78	(79)	80
81	82	(83)	84	85	86	87	88	(89)	90
91	92	93	94	95	96	(97)	98	99	100

Prime factors

Number	All factors	Prime factors
20	1, 2, 4, 5, 10, 20	2, 5
28	1, 2, 4, 7, 14, 28	2, 7
34	1, 2, 17, 34	2, 17

Odd one out

6, all the others are factors of 15.
40, all the others are factors of 27.
7, all the others are factors of 8.

Pages 8–9

Rounding 4-digit numbers ...

Nile: 6670 km, 6700 km, 7000 km
Mississippi: 6280 km, 6300 km, 6000 km
Ganges: 2530 km, 2500 km, 3000 km
Amazon: 6490 km, 6500 km, 6000 km
Congo: 4750 km, 4700 km, 5000 km

Rounding 6-digit numbers ...

Age UK: £115 000, £120 000, £100 000
Africa
Wildlife: £593 000, £590 000, £600 000
Homeless: £852 000, £850 000, £900 000
Water
Aid: £913 000, £910 000, £900 000

At the airport

1. a) 249 999 **b)** 150 000
2. a) 64 999 **b)** 55 000

Pages 10–11

Counting in powers of ten

1110, 2110, 3110, 4110, 5110, 6110
34 562, 44 562, 54 562, 64 562, 74 562, 84 562
131 782, 231 782, 331 782, 431 782, 531 782, 631 782
567 943, 577 943, 587 943, 597 943, 607 943, 617 943
867 101, 868 101, 869 101, 870 101, 871 101, 872 101

Negative and positive numbers

–12, –8, –4, 0, 4, 8, 12
–15, –10, –5, 0, 5, 10, 15
–11, –9, –7, –5, –3, –1, 1
–10, –8, –6, –4, –2, 0, 2

Bank account balance

£45

Pages 12–13

Place value

two tenths $\left(\frac{2}{10}\right)$, eight hundredths $\left(\frac{8}{100}\right)$, nine thousandths $\left(\frac{9}{1000}\right)$, 1 unit (1), one thousandth $\left(\frac{1}{1000}\right)$, five tenths $\left(\frac{5}{10}\right)$

Rounding

1. 85, 84.6 **2.** 85, 84.5 **3.** 84, 84.1
4. 83, 83.3 **5.** 83, 82.8 **6.** 83, 82.6
7. 82, 81.9 **8.** 81, 81.2

Decimal detective

4.28; Other numbers are any eight from: 4.25, 4.26, 4.27, 4.29, 4.30, 4.31 to 4.34

Pages 14–15

Adding large whole numbers

49 374, 17 631,
48 785, 643 734, 1 000 519

Subtracting large whole numbers

1893, 6932,
21 220, 2858, 3139

Missing digits

4567 + 56 410 = 60 977, 98 267 + 39 252 = 137 519, 79 658 – 5386 = 74 272

Pages 16–17

Multiplying by 10, 100 and 1000

240, 2400, 24 000
350, 3500, 35 000
6540, 65 400, 654 000
9010, 90 100, 901 000
87, 870, 8700
42, 420, 4200
12.4, 124, 1240

Dividing by 10, 100 and 1000

32.7, 3.27, 0.327
50.1, 5.01, 0.501
6.9, 0.69, 0.069
8.2, 0.82, 0.082
13.462, 1.3462, 0.13462
4.759, 0.4759, 0.04759
0.424, 0.0424, 0.00424

Maths chat

Tom is not correct. For example: 10 000 divided by 1000 = 10

Pages 18–19

Multiplying whole numbers

3804, 8073, 22 815, 133 798, 108 091

Multiplying decimals

607.6, 820, 5847.6, 71.67, 284.6

Game: Floor plan

Living room = 34.9 m^2, Dining room = 12.69 m^2, Play room = 8.46 m^2, Bathroom = 6.38 m^2, Kitchen = 40.64 m^2, Hallway = 18.68 m^2, Total area = 121.75 m^2

Multiplication puzzle

7.64 × 9 = 68.76; 6.79 × 4 = 27.16

Pages 20–21

Dividing larger numbers

$171 \div 6 = 28 \text{ r } 3;\ 28\frac{3}{6} = 28\frac{1}{2};\ 28.5$

$891 \div 5 = 178 \text{ r } 1;\ 178\frac{1}{5};\ 178.2$

$8455 \div 4 = 2113 \text{ r } 3;\ 2113\frac{3}{4};\ 2113.75$

Division problems

1. Each slice is 3.75 cm long
2. $141\frac{4}{6}$ g = $141\frac{2}{3}$ g per cake
3. a) Each friend gets 38 shells, with 2 shells left over. **b)** No, it does not make sense to use fractions or decimals in the answer since you would not break the shells.

Division: Round up or round down?

1. 2 buses **2.** 16 full boxes of eggs
3. 23 complete rows of tomato plants

Pages 22–23

Money problems

1. a) 2375 pennies **b)** £23.75 **2.** £299
3. 14 packets of crisps **4.** £2.20

Scale drawings

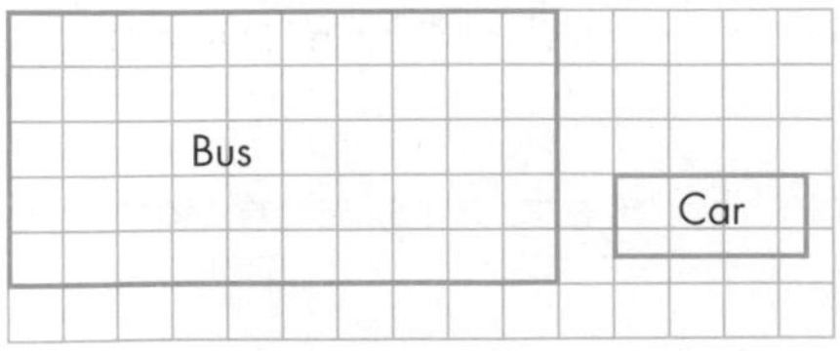

Bus
Car

Game: Decimal totals

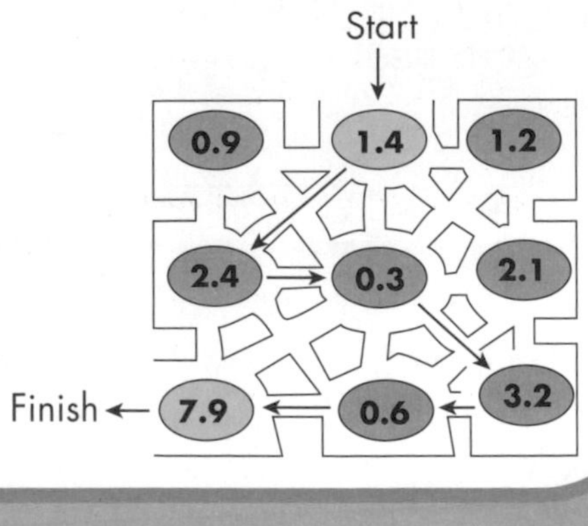

Start
0.9
1.4
1.2
2.4
0.3
2.1
Finish
7.9
0.6
3.2